Airplane

Ax

Apple

Acorn

A

Aquarium

Accordion

Ant

Ambulance

Bus

Bow tie

Beret

Briefcase

Beard

Baby bottle

Beads

Bb

Balloons

Bouquet

Blackberry

Bananas

Beet

Bear

Concrete mixer

Carrot

Crown

Cupcake

Candle

Car

Circle

Cherries

Chemise

Clown

Chamomile

Campfire

Dog

Duck

Doughnut

Dirigible

Dress

Dump truck

Drum

Dd

Door

Darts

F

FISH

G

GIRAFFE

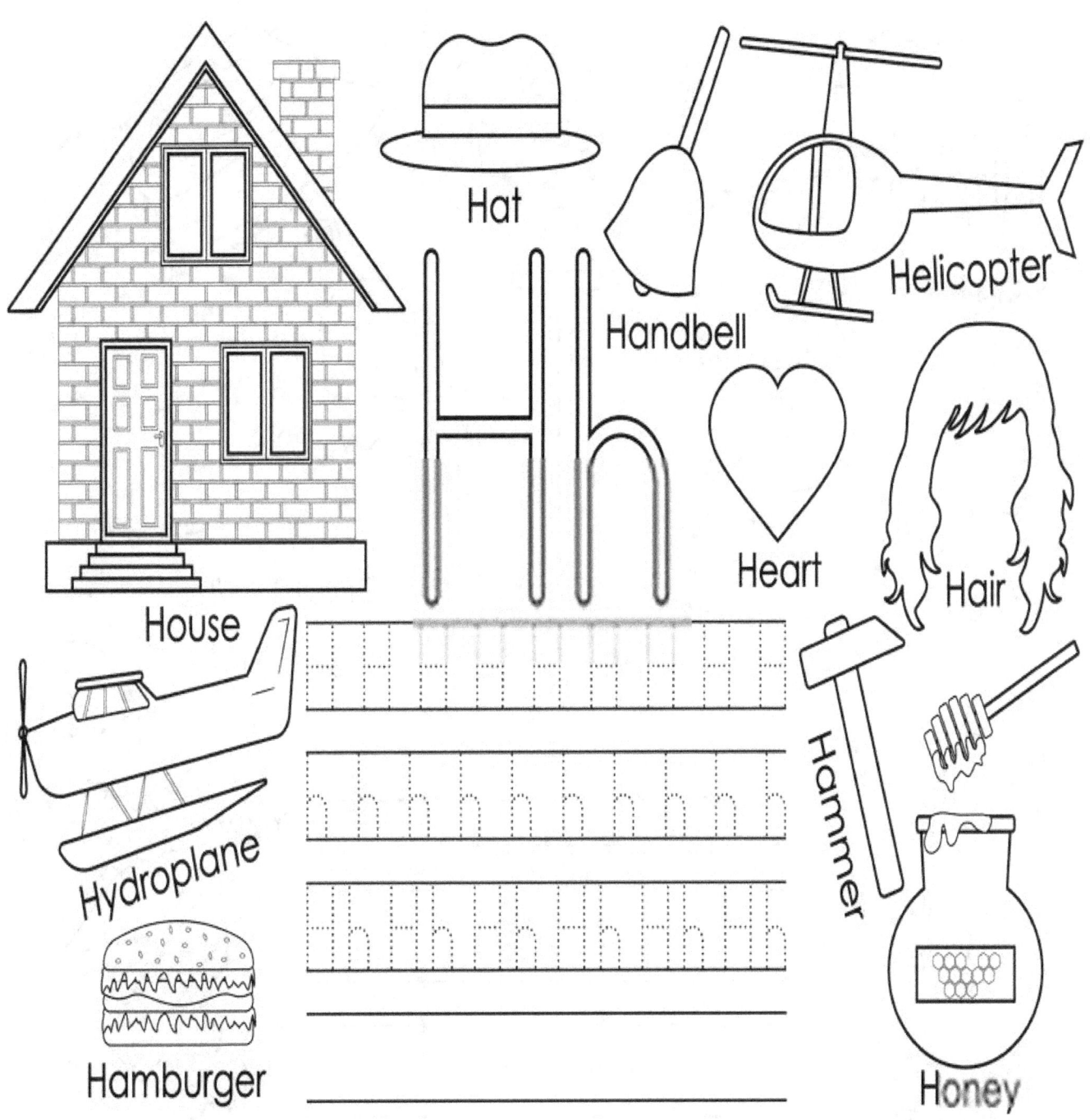

Hat

Handbell

Helicopter

Heart

Hair

House

Hydroplane

Hamburger

Hammer

Honey

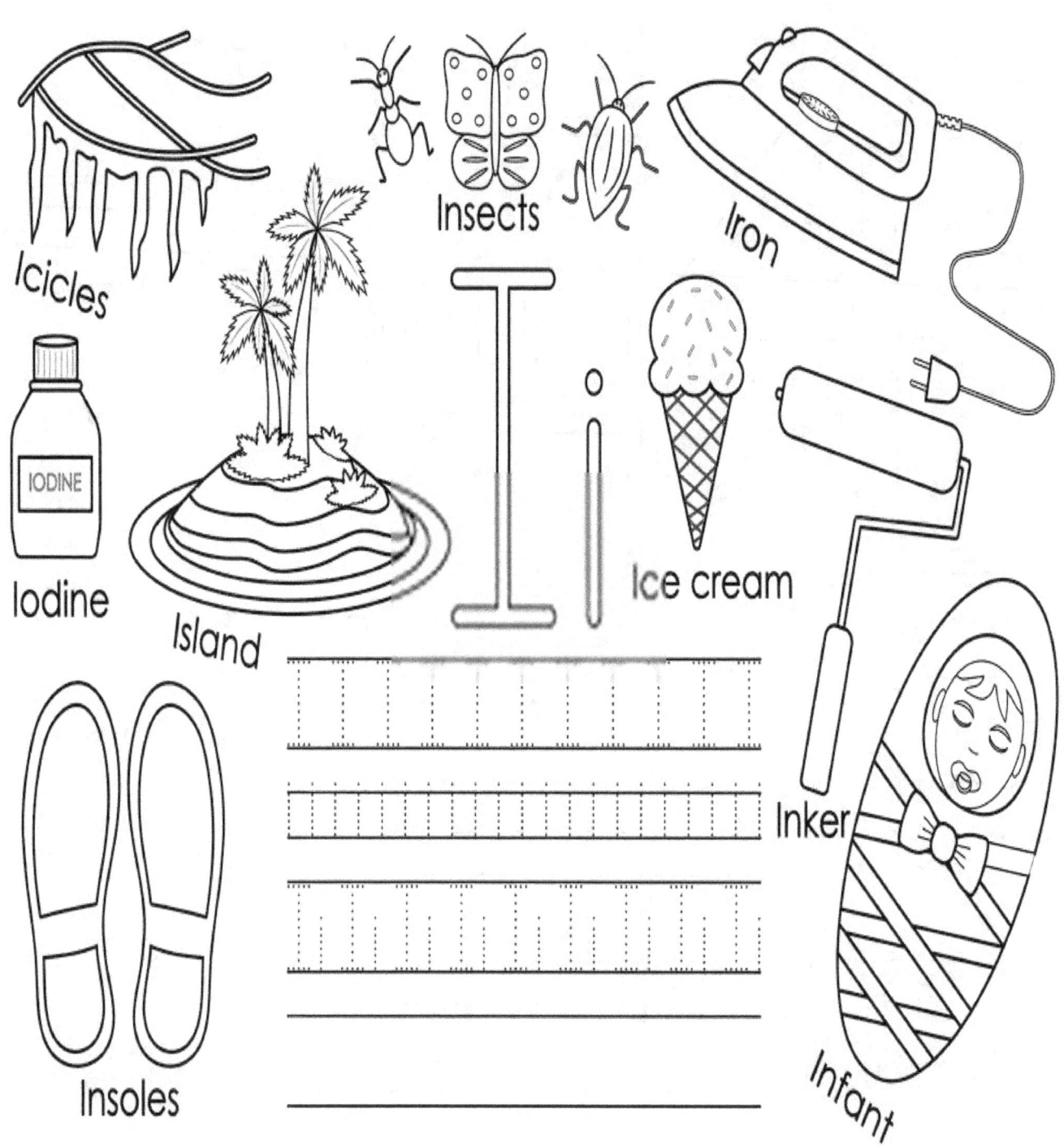

Icicles

Insects

Iron

Iodine

Island

Ice cream

Inker

Insoles

Infant

L

LIZARD

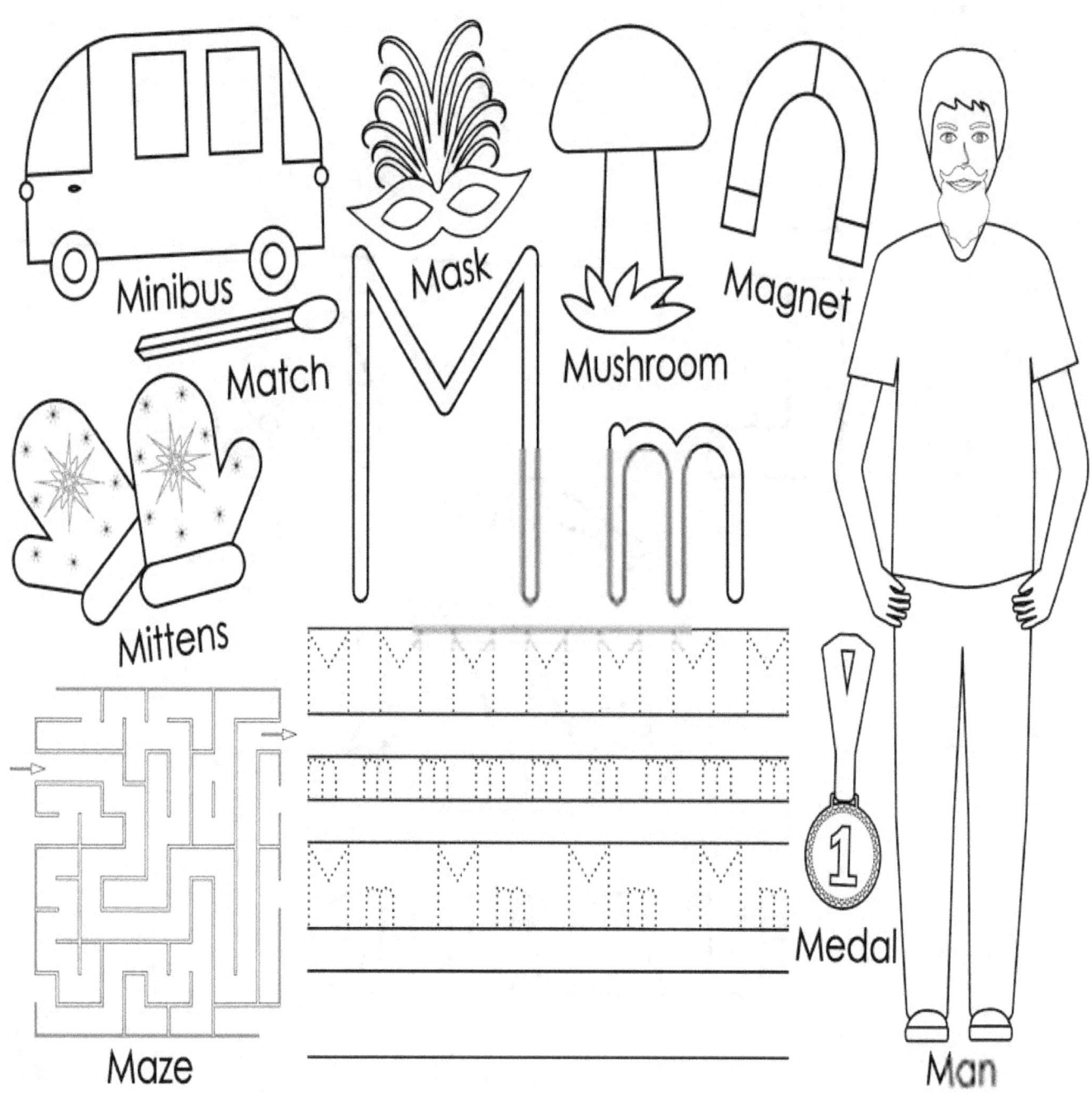

Minibus

Match

Mask

Mushroom

Magnet

Mittens

Mm

Maze

Medal

Man

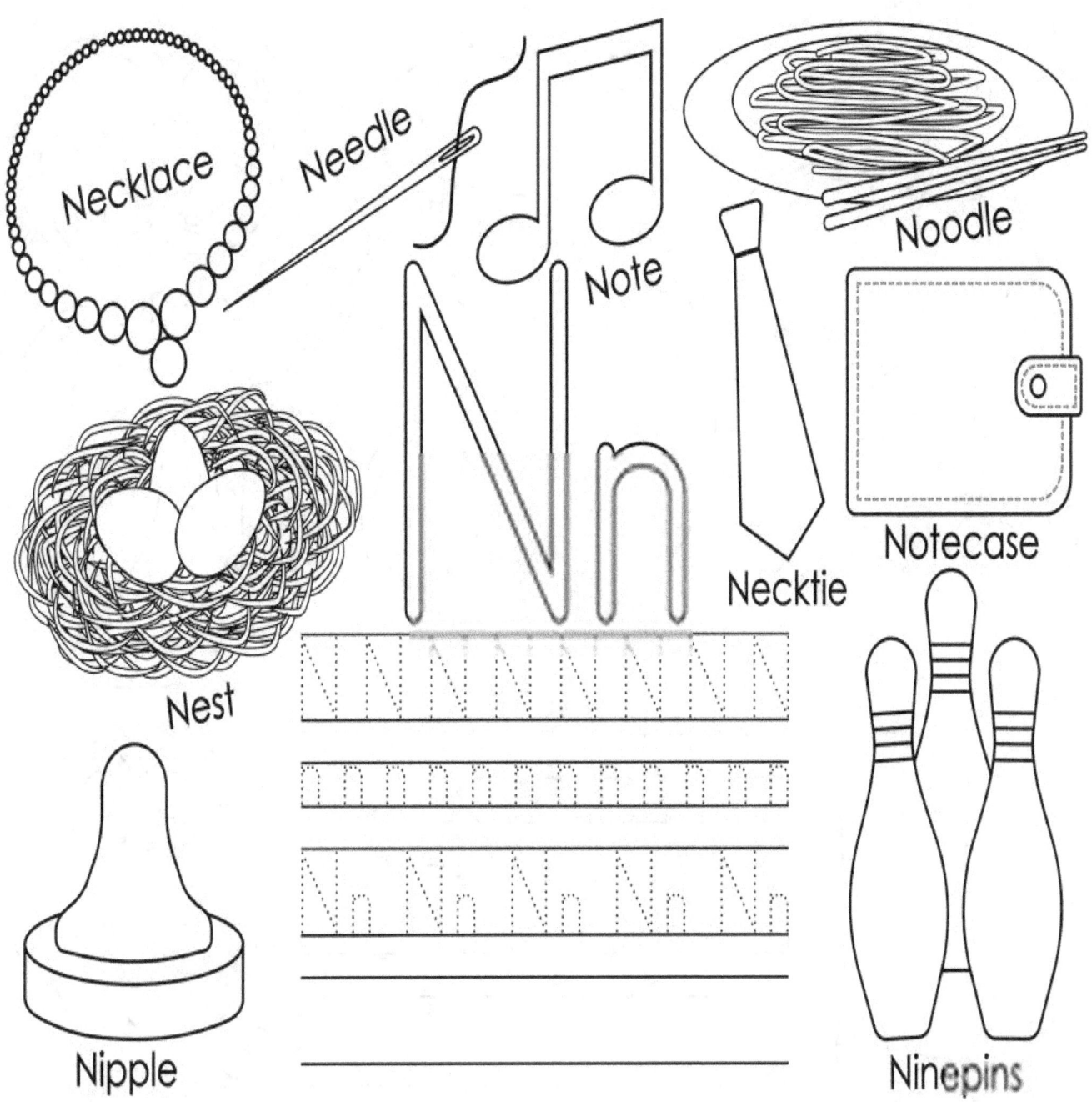

Necklace

Needle

Note

Noodle

Necktie

Notecase

Nest

Nipple

Ninepins

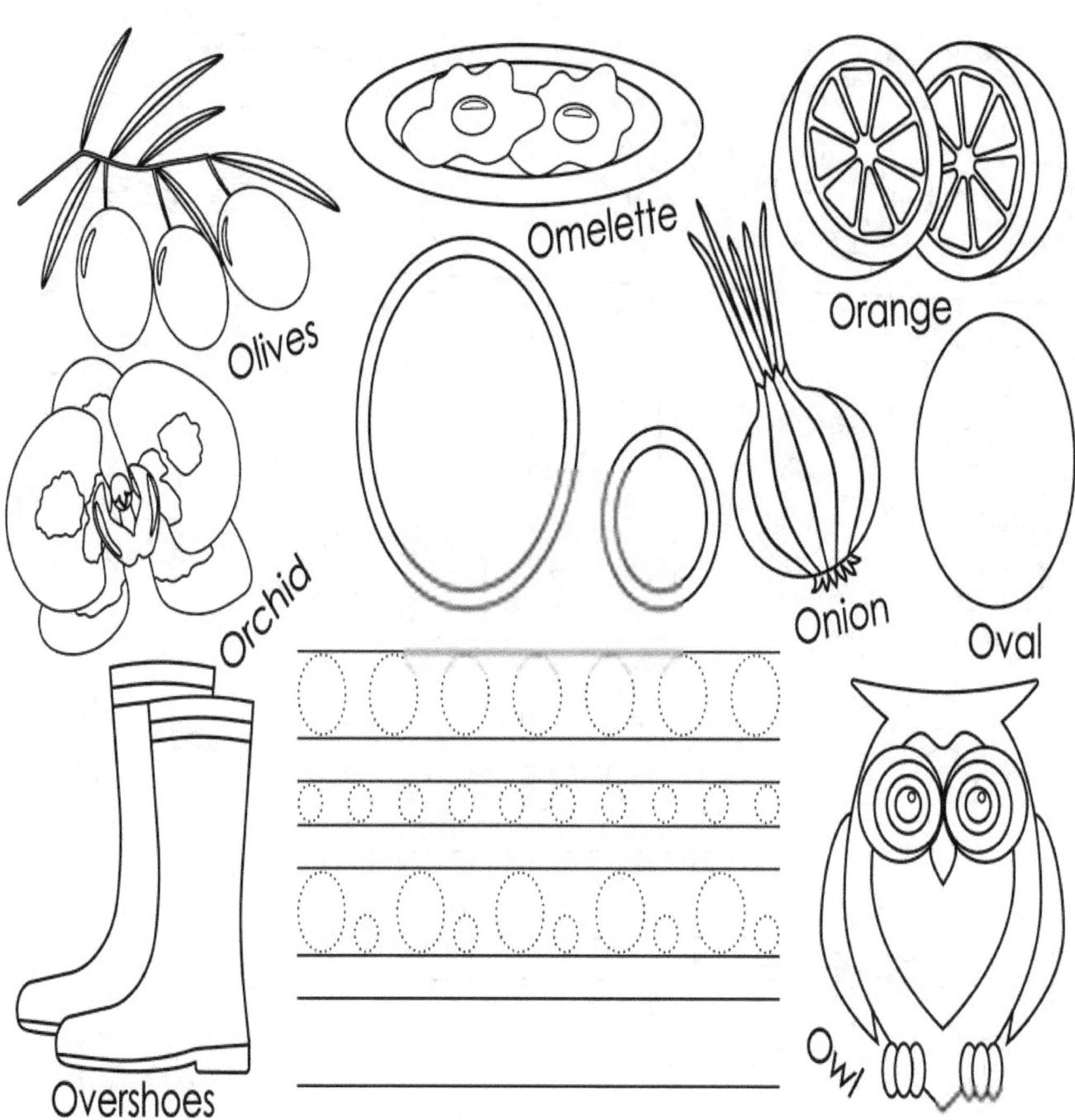

Olives

Omelette

Orange

Orchid

Onion

Oval

Overshoes

Owl

14

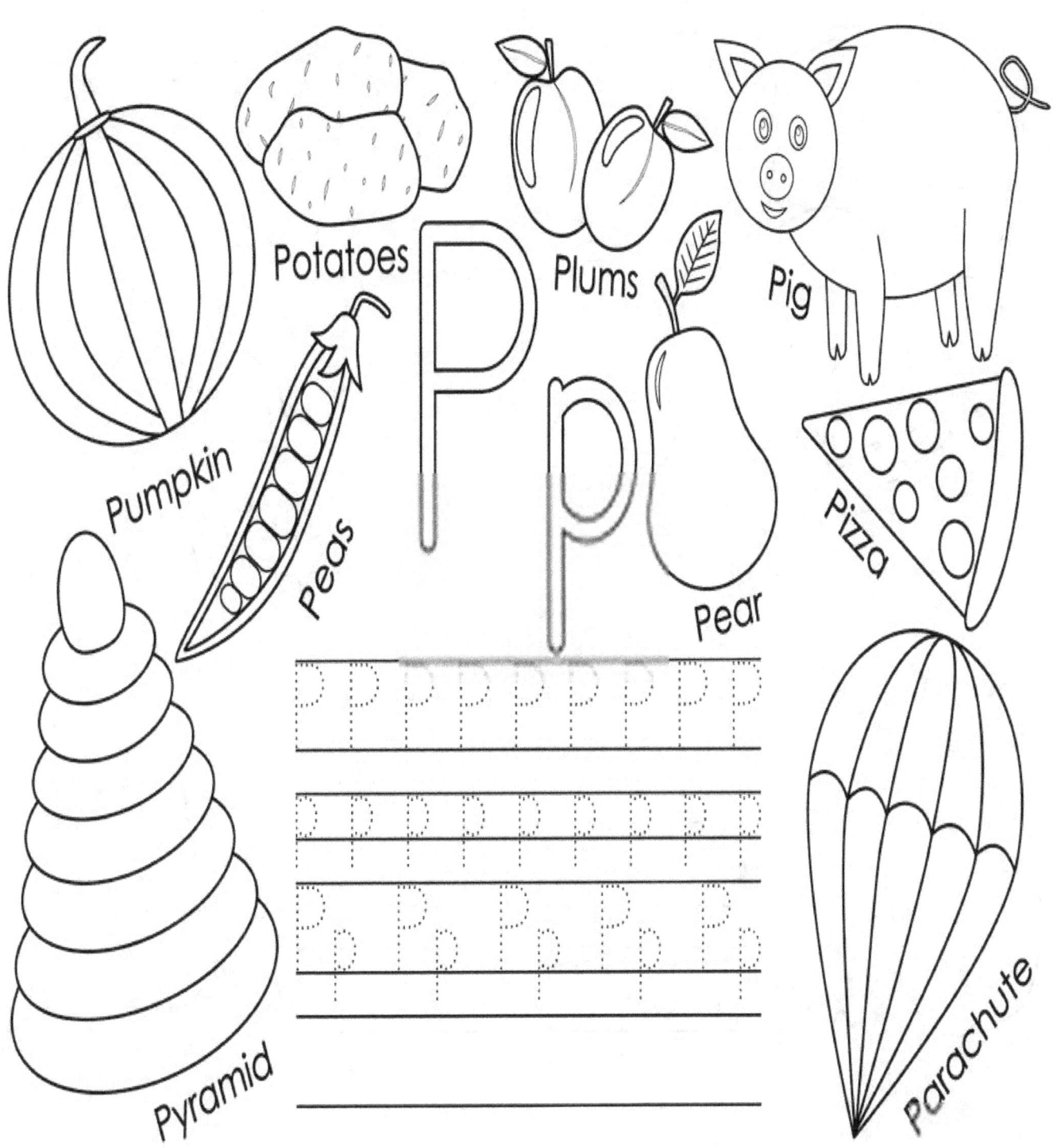

Potatoes

Plums

Pig

Pumpkin

Peas

Pear

Pizza

Pyramid

Parachute

15

Quilt

Quadrangles

Quill

Question mark

Quarter

Quinces

Queen

16

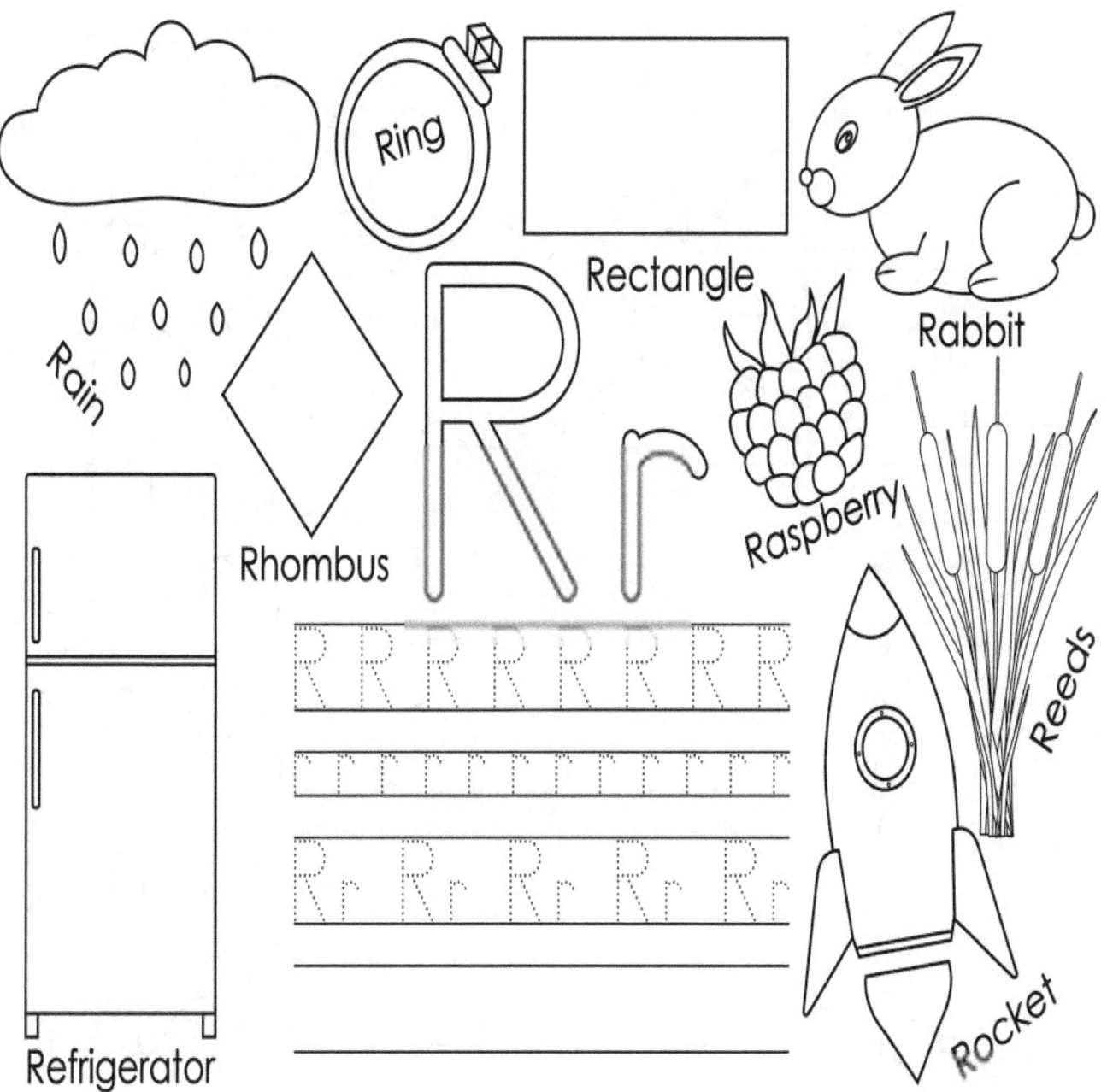

Rain

Ring

Rectangle

Rabbit

Rhombus

Raspberry

Refrigerator

Reeds

Rocket

17

Shed

Star

Sun

Sheep

Strawberry

Snowman

Sugar candy

Shirt

Snowflake

Skirt

18

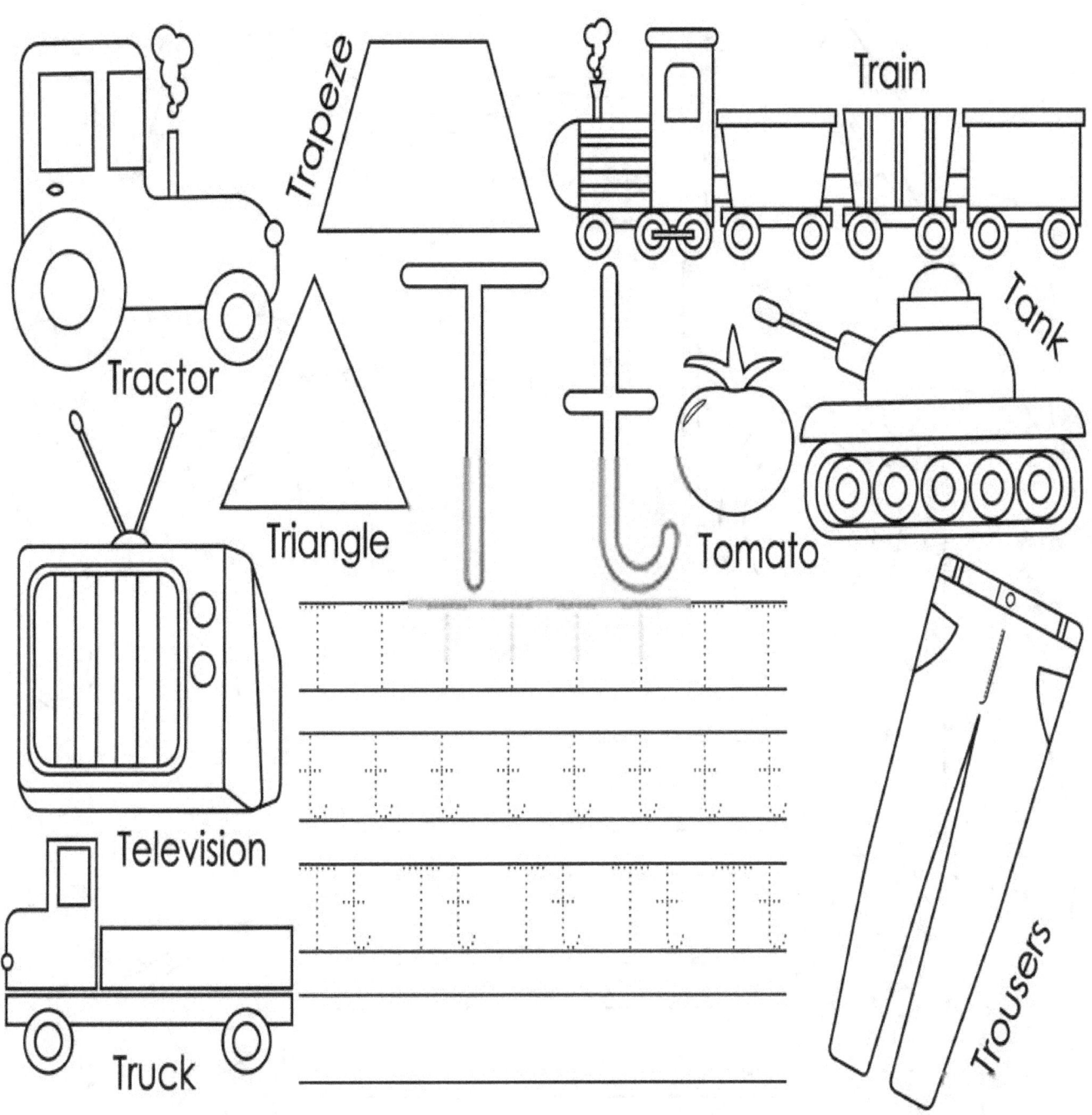

Tractor

Trapeze

Train

Triangle

Tank

Tomato

Television

Truck

Trousers

19

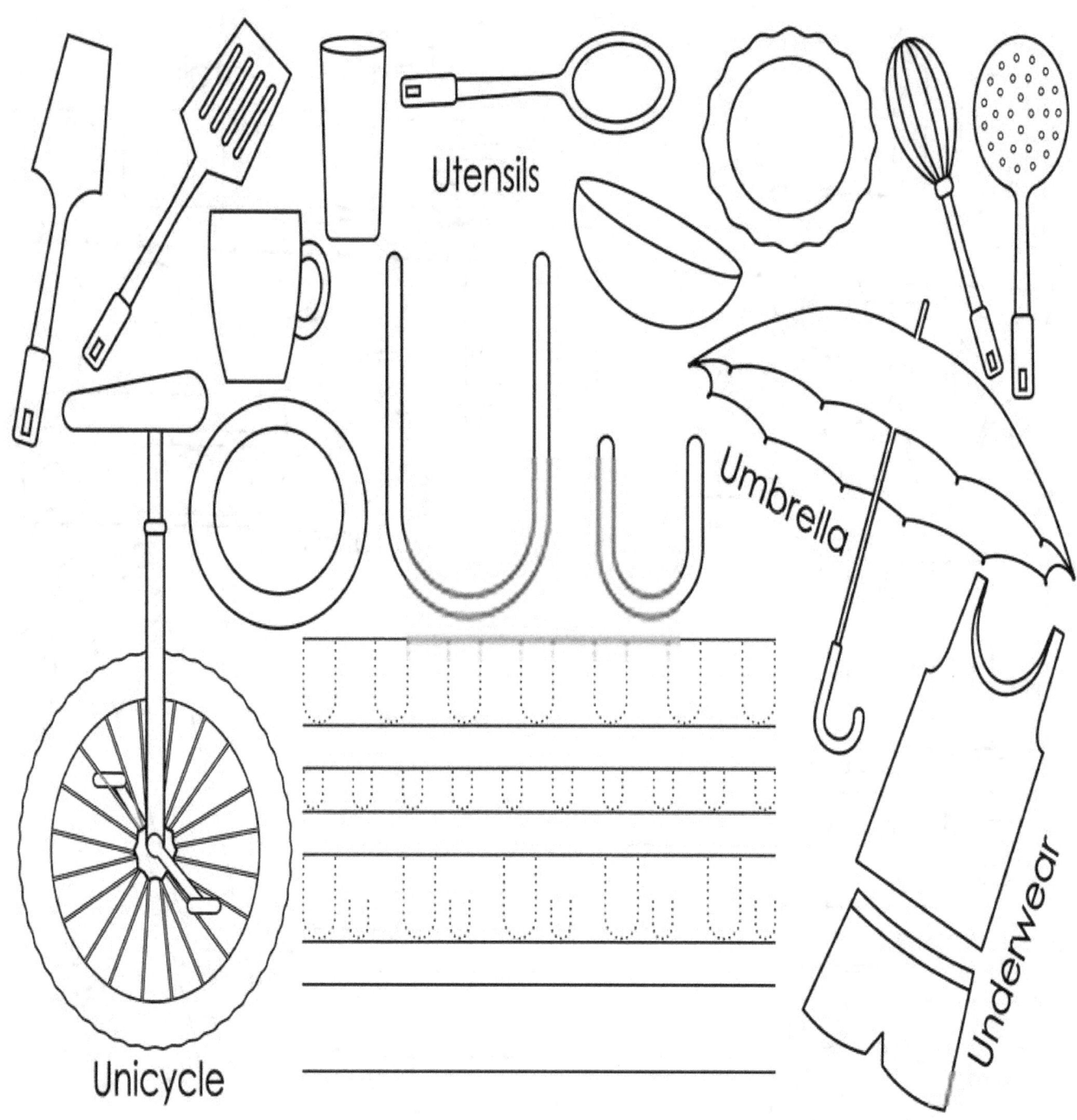

Utensils

Umbrella

Underwear

Unicycle

20

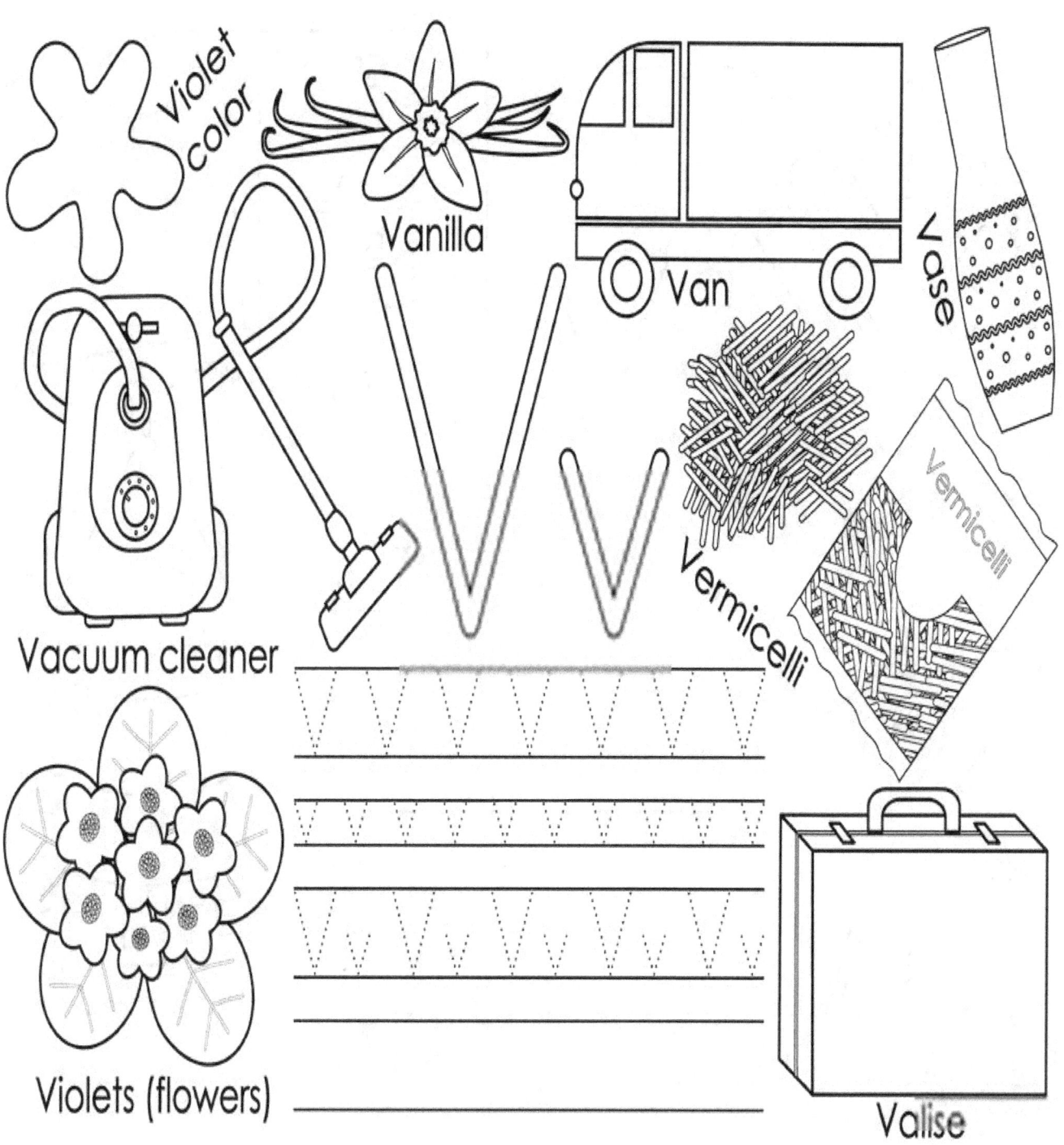

Violet color

Vanilla

Van

Vase

Vacuum cleaner

Vermicelli

Violets (flowers)

Valise

21

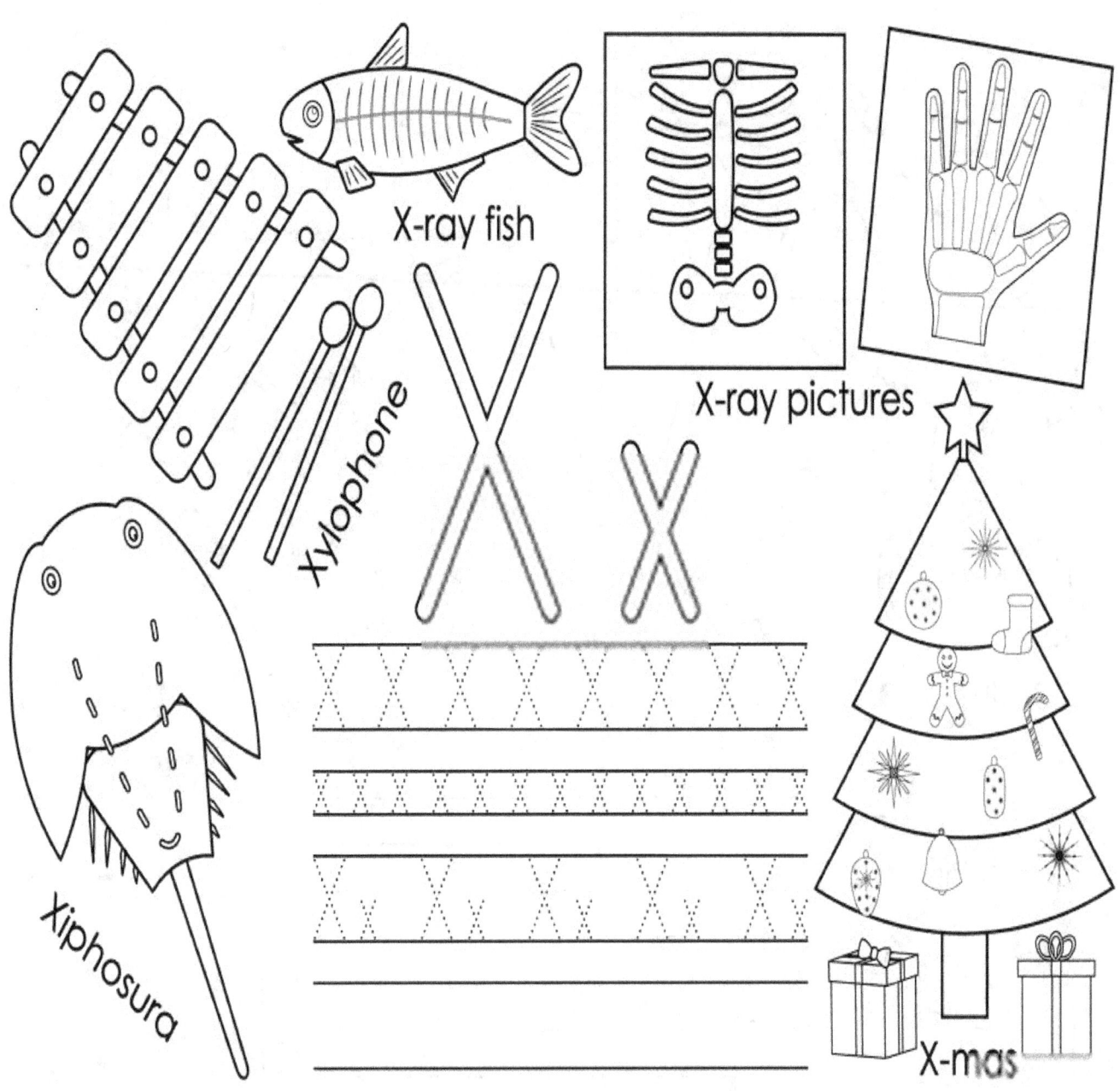

X-ray fish

X-ray pictures

Xylophone

Xiphosura

X-mas

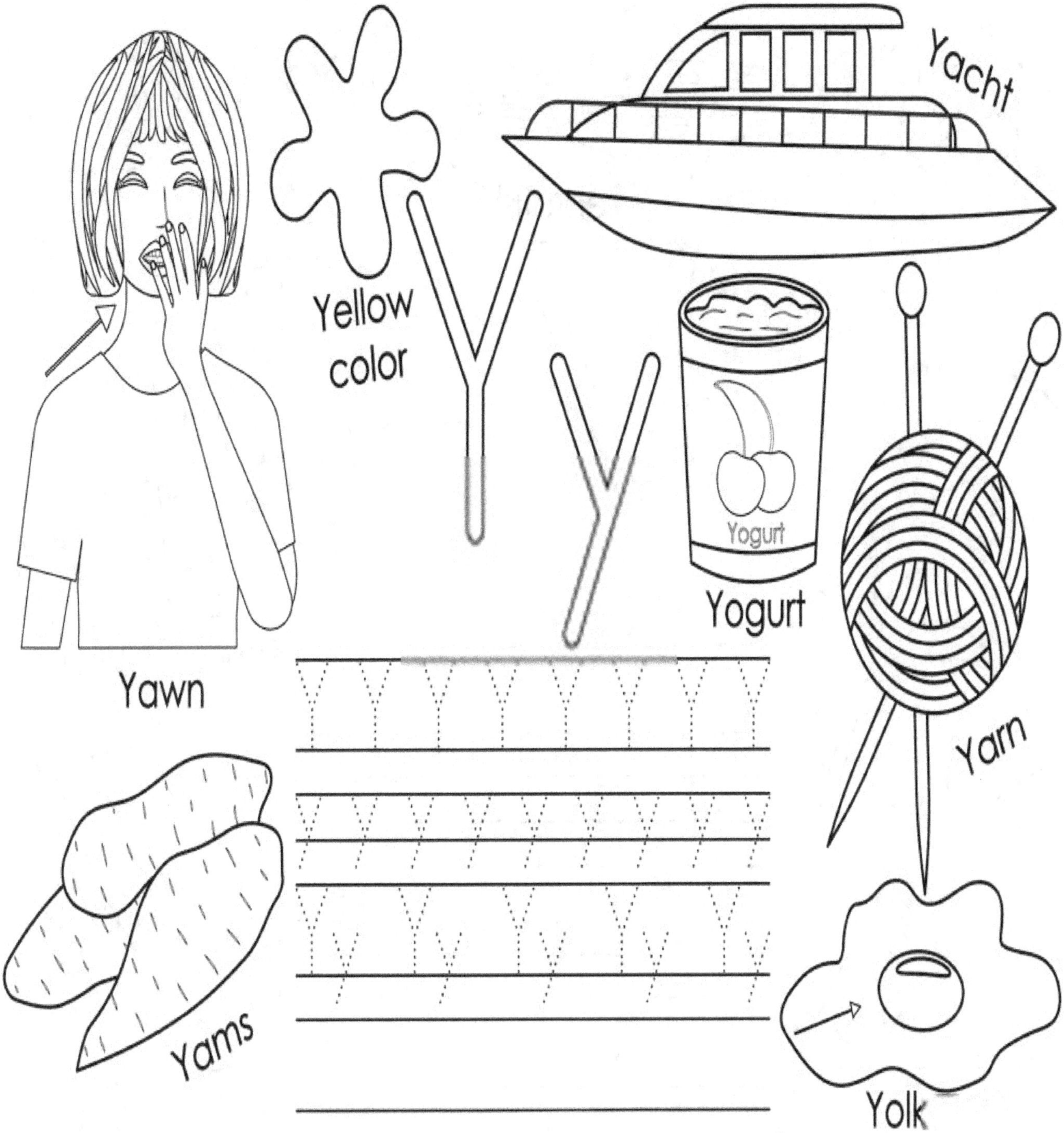

Yacht

Yellow
color

Yogurt

Yarn

Yawn

Yams

Yolk

23

Zebrafish

Zigzag

Zinnia

Zero

Zebra

Zipper

Zucchini

E

ELEPHANT

Z

ZEBRA

H

HORSE

L

LIZARD

S

SHARK

U

URIAL

D

DOG

X

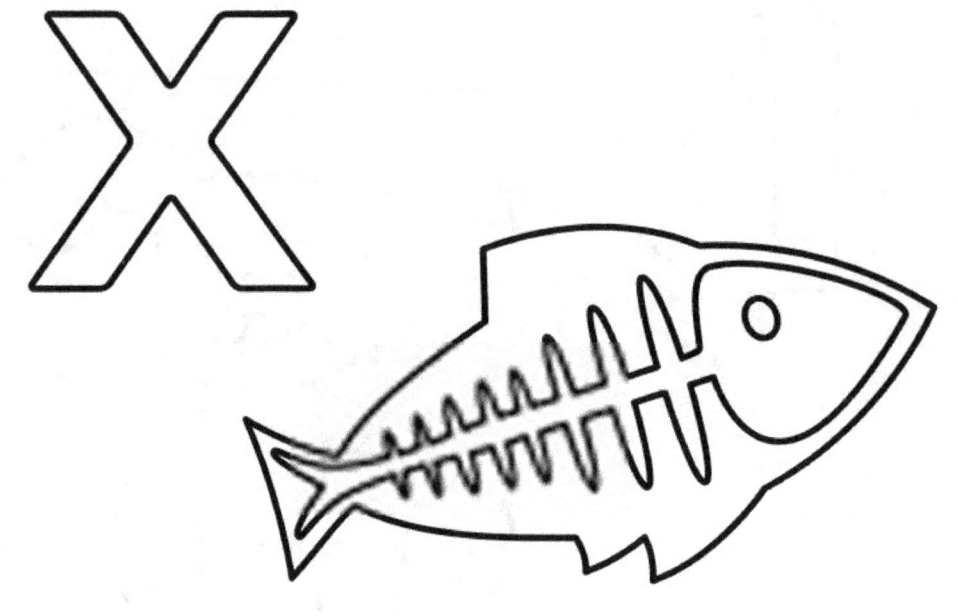

X-RAY FISH

P

PARROT

www.ingramcontent.com/pod-product-compliance
Lightning Source LLC
Chambersburg PA
CBHW081708220526

45466CB00009B/2912